Water Footprint Assessment

A Guide for Businesses

T0298876

David Tickner

Chief Freshwater Adviser, WWF-UK,
Living Planet Centre, Rufford House,
Brewery Road, Woking GU21 4LL, UK
dtickner@wwf.org.uk

Ashok Kumar Chapagain

Science Director, Water Footprint Network,
International Water House, The Hague,
The Netherlands
ashok.chapagain@waterfootprint.org

First published in 2015 by Dō Sustainability

87 Lonsdale Road, Oxford OX2 7ET, UK

ISBN 978-1-910174-57-9 (eBook-ePub)

ISBN 978-1-910174-58-6 (eBook-PDF)

ISBN 978-1-910174-56-2 (Paperback)

A catalogue record for this title is available from the British Library.

Dō Sustainability strives for net positive social and environmental impact. See our sustainability policy at **www.dosustainability.com**.

Page design and typesetting by Alison Rayner

Cover by Orest Viruta

For further information on Dō Sustainability, visit our website: **www.dosustainability.com**

DōShorts

Dō Sustainability is the publisher of DōShorts: short, high-value business guides that distill sustainability best practice and business insights for busy, results-driven professionals. Each DōShort can be read in 90 minutes.

New and forthcoming DōShorts – stay up to date

We publish new DōShorts each month. The best way to keep up to date? Sign up to our short, monthly newsletter. Go to **www.dosustainability.com/newsletter** to sign up to the Dō Newsletter. Some of our latest and forthcoming titles include:

- *Creating Employee Champions: How to Drive Business Success Through Sustainability Engagement Training*   Joanna M. Sullivan
- *Smart Engagement: Why, What, Who and How*   John Aston & Alan Knight
- *How to Produce a Sustainability Report*   Kye Gbangbola & Nicole Lawler
- *Strategic Sustainable Procurement: An Overview of Law and Best Practice for the Public and Private Sectors*   Colleen Theron & Malcolm Dowden
- *The Reputation Risk Handbook: Surviving and Thriving in the Age of Hyper-Transparency*   Andrea Bonime-Blanc
- *Business Strategy for Water Challenges: From Risk to Opportunity* Stuart Orr and Guy Pegram
- *Accelerating Sustainability Using the 80/20 Rule*   Gareth Kane
- *The Guide to the Circular Economy: Capturing Value and Managing Material Risk*   Dustin Benton, Jonny Hazell and Julie Hill

- *PR 2.0: How Digital Media Can Help You Build a Sustainable Brand*   John Friedman

- *Valuing Natural Capital: Futureproofing Business and Finance*   Dorothy Maxwell

- *Storytelling for Sustainability: Deepening the Case for Change*   Jeff Leinaweaver

- *Beyond Certification*   Scott Poynton

- *21st Century Growth: Beyond the Water-Energy-Food Nexus*   Will Sarni

- *Creating a Culture of Integrity: Business Ethics for the 21st Century*   Andrea Spencer-Cooke & Fran van Dijk

Subscriptions

In addition to individual sales of our ebooks, we now offer subscriptions. Access 60+ ebooks for the price of 6 with a personal subscription to our full e-library. Institutional subscriptions are also available for your staff or students. Visit **www.dosustainability.com/books/subscriptions** or email **veruschka@dosustainability.com**

Write for us, or suggest a DōShort

Please visit **www.dosustainability.com** for our full publishing programme. If you don't find what you need, write for us! Or suggest a DōShort on our website. We look forward to hearing from you.

Abstract

WATER IS ONE OF THE MOST PRESSING and complex issues facing sustainability practitioners. As population growth, economic development and climate change intensify pressures on rivers, lakes and aquifers, many companies are becoming aware that water scarcity and pollution present strategic risks to their operations, supply chains and reputations. They seek practical tools which can help them to understand and assess levels of risk. Water Footprint Assessment tools developed primarily by the research sector are gaining increasing attention in this context. However, there is debate among experts and non-experts about the merits of Water Footprint Assessment. Based on practical experience and case studies, the book will explain the issues and provide guidance for companies on the pros and cons of using Water Footprint Assessment and similar approaches.

....

About the Authors

DAVID TICKNER is Chief Freshwater Adviser at WWF-UK. He has spent the last 15 years providing strategic leadership to river conservation projects around the globe and advising governments and companies on water management practice and policy. Dave is currently also a Research Fellow at the University of East Anglia and an Associate Editor of the journal *Frontiers in Freshwater Science*. Previously, Dave worked in the UK government's environment ministry and led WWF's conservation and policy programme for the River Danube. He was also a founding non-executive director of a not-for-profit company, Water and Sanitation for the Urban Poor (WSUP), and an advisor to the group environment committee of Standard Chartered bank. Dave holds a PhD in hydro-ecology and has authored, edited or contributed to a range of popular articles, scientific papers, blogs and books on water-related issues. You can follow him on Twitter (**@david_tickner**).

ASHOK CHAPAGAIN is Science Director at the Water Footprint Network (WFN), where he is responsible for developing and implementing WFN's research programme. He leads WFN's training and other knowledge-sharing activities and helps multinationals to understand global water risks and supports them to formulate appropriate response strategies based on Water Footprint Assessment. Previously,

ABOUT THE AUTHORS

Ashok worked for WWF-UK for six years, developing policy and practice on water stewardship and water security. Through his PhD research in the field of water systems and policy analysis, Ashok co-developed the concept of Water Footprint Assessment and co-authored the *Water Footprint Assessment Manual*, used globally as the standard in this field. He has led the development of various tools related to water accounting. His list of publication includes three books and more than 50 articles in peer-reviewed scientific journals, conference proceedings, book chapters and reports. You can follow him on Twitter (**@A_Chapagain**).

Acknowledgments

THE AUTHORS' EXPERIENCE of Water Footprint Assessment has mostly come about through our day jobs doing research and gaining practical experience in WWF-UK, University of Twente, UNESCO-IHE and the Water Footprint Network. We'd like to thank all the colleagues who have helped us over this period.

Part of the content of this paper has been adapted from an earlier academic paper: Chapagain, A. K. and Tickner, D. 2012. Water footprint: Help or hindrance? *Water Alternatives* (Volume 5): 563–581.

Contents

CHAPTER 1

Introduction

IMAGINE THAT YOU NEED TO BUY SOME BREAD. As someone who thinks about sustainability (which we assume is the case given that you're reading this DōShort), you want to buy the loaf with the best environmental performance. You know that your decision should take account of several factors. One of these is water.

In the shop you pick up two loaves to compare. The packaging of each declares its Water Footprint. One has a footprint of 10 litres. For the other it's 100 litres. There's not much different in price or taste. So you decide that the best choice is the bread with the lowest footprint. Right?

Not necessarily.

It could be the case that the wheat that made the flour for the 10 litre loaf was grown in an irrigated system using water abstracted from streams in parched Australia. The flour for the 100 litre bread might have come from the rain-fed fields in relatively soggy Scotland. Now ask yourself, which loaf has the greater environmental impact? More specifically, which bread should you buy if you care about our over-stretched river and wetland ecosystems?

If you want to address your carbon footprint you need to know your numbers. The more CO_2 or other greenhouse gas you pump into the atmosphere, the greater will be your contribution to climate change. You can tell if you're reducing your negative impacts because your footprint will diminish. There are implementation challenges and methodological issues to grapple with, but the concept is simple. And the idea is the same wherever you are in the world.

Water is more complex. Size may not matter so much; context is often more important. Knowing that the production of a loaf of bread requires 10 or 100 litres of water might be a useful starting point but if you want to understand how sustainable your Water Footprint is you need more information – about the degree of water scarcity where the wheat was grown, the pollution levels in rivers or aquifers, the extent to which multiple users in that location are collectively putting pressure on hydrological resources and the social costs and benefits of using all that H_2O.

So what is the point of estimating Water Footprint numbers if they don't provide us with a simple guide to action?

This DōShort will try to answer that question. It describes the evolution of the concepts that underpin Water Footprint tools and methods and it explains the ways in which Water Footprints can be calculated. It discusses the way in which Water Footprints have been used to provide answers to questions that many businesses are asking and it highlights some of the questions that Water Footprint approaches can't answer. And it sets out the main advantages and disadvantages – the positives and pitfalls – of using Water Footprint approaches.

Perhaps most importantly, this book provides some guiding principles – or golden rules – that you should follow if you're thinking about looking at Water Footprints as a way of helping your business.

First, let's remind ourselves of what's happening to the world's water resources and why all of this matters . . .

CHAPTER 2

Context: Water, Business and Risk

2.1 **The water's edge**

WE ALL LIVE AT THE WATER'S EDGE. Whether we're at the end of a pipe or on the banks of a river, we rely on water to grow the food we eat, generate our energy and lubricate the industries which provide the stuff we buy. Water also provides for many of our most basic human needs. Contrary to popular perception, the water that we use doesn't really come from the tap or tank or pipe. We take our water from nature – from rainfall or rivers, from lakes or aquifers. After we've finished with it, we return some of our water, often in a reduced or polluted state, to those sources.

There is increasing concern that the world is facing a water crisis,[1] driven by global 'megatrends' including rapid population growth, epochal shifts in the global economy and the looming impacts of climate change. All of these factors will add to the pressures on the rivers, lakes, aquifers and wetlands from which much of our water is taken.

The consequences of such pressures are already being felt by people and wildlife across the planet and unless governments and others implement rapid and radical improvements in water management the situation is likely to get worse. Water resources are being consumed faster than they can be replenished in large parts of China, India, Mexico, the Middle East, the Mediterranean region, Central Asia, Australia, southern Africa and the USA. Stories about conflicts between water users arise with depressing frequency.[2] The impacts of pollution, chronic water scarcity and acute drought in some countries have impacted on households, towns and cities. These changes have also influenced global wholesale and retail prices of commodities (even as we are writing this DōShort, a drought in Brazil is reported to have driven the costs of coffee to record levels[3] and there is mounting concern about the economic impacts of an historic drought in California). Climate change introduces a new element of unpredictability to rainfall and river flows which is very likely to exacerbate matters.

Ecosystems are suffering too. The combination of overuse of water, pollution, a huge expansion of water infrastructure such as dams, overfishing and the introduction of invasive species is profoundly altering the health of river and wetland habitats. As one indicator of this, the Yangtze River dolphin – the *baiji* in Chinese – is thought to be the first cetacean (whale or dolphin) to be driven to extinction by humans. Dolphin species in the Ganges, Indus and Mekong Rivers are faring little better. It's not just dolphins: authoritative reports[4] have described rapid declines in a wide variety of fish, mammals, birds, amphibians and other creatures which live in, or depend upon, freshwater.

2.2 The good news?

It's not all doom and gloom. In global terms there is probably enough water to go around, despite those megatrends. Even in countries or regions where water is scarce, innovations in farming practice and better water allocation rules can make a difference. We know from experience that rivers which have been grossly polluted by sewage and industrial effluents can be cleaned up relatively quickly given sufficient political impetus and capital investment (tackling pollution from other sources, such as agriculture, is trickier). There's hope for freshwater biodiversity too. In the parts of Europe, for instance, populations of some wetland creatures, such as otters and bitterns, are recovering after decades of decline.

Most governments understand that water is important even if many have struggled to address the conflicting demands from an increasing number of water users. Some – especially those which have suffered crises through repeated floods or droughts or other disasters – have begun to invest financial and political capital in efforts to tackle the problems. China provides an interesting example. After the Yellow River dried up once too often in the late 1990s, the Beijing authorities put in place radical measures to restore its flow. More recently, the Chinese government made water conservancy the central theme of its 'number one' policy in 2011 and set out 'red lines' on water pollution, water allocations and efficiency of water use. It also announced a massive budget hike for water management. All of this has been prompted as much by concerns about the political and economic consequences of scarcity, pollution and ecosystem degradation as by any altruistic concern for biodiversity.

Other countries have also responded to chronic water challenges. Since the mid-1990s a wave of progressive water policies has swept the planet. New laws in countries such as Mexico, Kenya and South Africa put the highest priority for water allocation on meeting the most basic human needs and on ensuring that a strategic reserve of water flows through river systems to downstream users (hydrologists and freshwater ecologists call this an 'environmental flow'). Withdrawals of water for economically productive purposes – such as agriculture or industry – are meant to be secondary priorities.

There's a catch, inevitably. The combination of massive demand, economic dependence on water, contested politics, weak institutional arrangements and a complex array of vested interests has meant that it has proven very difficult for governments to implement such policies. Even in China, the Yellow River isn't yet flowing to the levels that were intended despite the clear direction from Beijing. Water management bodies are insufficiently resourced in many countries and when it comes to the crunch of re-allocating water from one user to another or enforcing stricter pollution controls they frequently find that the political wind is against them. Strategically important sectors, such as energy utilities, or groups with strong influence, such as farmers, often receive the water they ask for regardless of what the policy says.

2.3 **Risky business**

Think of something you bought recently – anything will do. That loaf of bread we discussed earlier perhaps, or a new shirt, or the laptop or tablet you're using to read this. It's very likely that water has been required for irrigation, processing, cleaning or cooling purposes at some point in its production. In most instances there was no alternative substance that could have been used. Water is both universally needed and non-substitutable.

Think for a moment about what this could mean for your business. You might have operations or, more likely, suppliers scattered across continents. For many, water will be a critical resource. It may be the case that, in one or more of those locations, water scarcity or pollution might be having an effect on production. If that's not already happening, it's increasingly likely that in future, somewhere in the world, there's a water-related risk which could affect your company.

Now apply this at the scale of whole economies. The combination of increased demand for water, pollution, poor management and climate change, together with ever-more complex and globalised supply chains, means that water-related risks arising in different parts of the planet are affecting more and more companies. Captains of industry are worrying. The World Economic Forum, in its 2015 survey of global business risks,[5] concluded that water crises are likely to have the biggest impacts on economies around the world, ahead of challenges such as fiscal crises in key economies and structural unemployment.

In recent years, experts and commentators have published hundreds, if not thousands, of books and articles about the impacts of water scarcity or pollution on companies. An accepted typology of water-related business risk has emerged (Figure 1, overleaf).[6] In essence, problems can stem from:

- Shifts in *regulation* such as the licensing and pricing of water abstractions or pollution permits. As well as any direct impacts on costs from new regulations, the uncertainty which arises from varying regulatory regimes can hamper business planning.

- Potential damage to *reputation* and brand value, stemming from changes in the public or stakeholder perception of the company's impacts on social values linked to water. Companies, especially multinationals with a high profile, can be easy, and sometimes fair, targets for those seeking to apportion blame for pollution or depletion of rivers, lakes and aquifers or problems linked to unequal access to water supplies.

- A straightforward *physical scarcity* of water, or at least a scarcity of water that meets quality standards necessary for production. Such scarcity can lead to factories or facilities shutting down, temporarily or permanently. This is, at root, the issue which gives rise to much regulatory or reputational risk, but it can in itself have direct impacts on a business, especially during drought conditions. (It's worth noting that in some places, flooding can pose problems for companies too.)

FIGURE 1. Water-related risks from a business perspective.

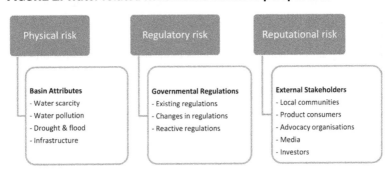

For many companies, these risks pose very immediate challenges. The costs of mitigation measures can be significant, in terms of financial or social capital, or both. Drinks manufacturers have been the first to feel the hydrological squeeze. If you talk to representatives from any well-known multinational beverage company, it's likely that they will regale you with stories of a bottling plant being closed because of concern about its real or perceived impact on local aquifers, or of a brewery needing to buy in tanker-loads of water because of shortfalls in piped supplies. In cities such as Dar es Salaam in Tanzania, companies like SABMiller (one of the world's largest brewers) have felt the need to invest substantial sums drilling new boreholes because public water utilities just haven't been able to cope with the burgeoning demands of a rapidly growing population.

Increasingly, voices from other sectors can be heard echoing these experiences. Nestlé (which makes food as well as drinks) is now vocal about the way in which water scarcity and pollution present strategic risks to agricultural supply chains. Retailers like Sainsbury's and M&S include water prominently among their sustainability priorities. IPIECA, the environmental body for the oil and gas sector, has launched an initiative to help its members understand and address water-related risks. Textiles and mining companies are, to one degree or another, following suit.

Major institutional investors are also now asking questions about water. CDP (formerly the Carbon Disclosure Project) surveys companies about water-related risks in order to inform investors of the state of play. Currently, the survey is backed by 573 different investors representing some US$60 trillion in assets. In CDP's 2013 survey, 70% of Global 500 companies reported exposure to substantive water-related risks; and 64% of them said that such risks were expected to impact now or in the next five years.[7]

2.4 What's yours is mine

As the global debate about water and business has grown, the concept of shared risk has emerged. Water is a common pool resource. Where pollution or scarcity occurs, there are normally multiple users or polluters of the river rather than a single culprit. Declining quantity or quality of water in any river, lake or aquifer will impact in some way on everyone who uses that water, as well as on freshwater ecosystems. The corollary is that, in any given location, a single company can't possibly manage its risks by acting alone.

To give a hypothetical example, a brewery manager who draws water from a particular river might put in place measures to ensure that his or her operation is supremely efficient in its use of water. But if all the other farmers, factories and facilities along the river remain profligate then the river might still dry up. Our brewery manager will then find that he or she still struggles to get enough water to meet production targets. The only way to ensure that water supplies keep flowing might be to work with neighbouring factory and facility managers to understand collective exposure to water-related risk and to encourage all of them to implement carefully defined water use standards. Other stakeholders – be they householders, dam operators, fisher-folk or conservationists – might also demand a say in the process.

All these actors might need to commit to a package of measures which could include, say, developing new water sources, stricter wastewater treatment, improving management of wetlands and economic measures such as water pricing. Because there are always trade-offs, and because water is a common pool resource and often regarded as a public good, governments or regulators will normally have the final say on standards and measures. Businesses can play an important role by drawing attention to the issues, encouraging transparent dialogue, implementing good water stewardship practice[8] and providing resources for some 'public good' measures such as enhanced river and aquifer monitoring.

2.5 **From confusion comes wisdom**

Water management is often complex, especially when it comes to designing and implementing standards and measures that address shared risks. It's not simple to balance upstream versus downstream users, the costs and benefits of different management options, short-term and long-term objectives, the needs of rich and poor people. For anyone who wants to engage with water, a first step is to understand that it can be a confusing issue. Silver bullets are rare.

If governments are to put in place targeted measures to ensure that rivers flow, aquifers stay clean and the wider population benefits, and if companies are to thrive in this context, they will need tools that can help them cut through the confusion, improve their understanding of shared risks and point to joint solutions. There has been substantial innovation in this field over the last decade, led by academia, NGOs and some of the most progressive (and vulnerable) multinational companies.

Among the new approaches which have emerged are Water Footprint Assessment techniques which map and measure the invisible, or virtual, link between water users in one part of the world, the products they grow or make and the companies and consumers, often in far off lands, who rely on such production.

CHAPTER 3

Water Footprint:
Concepts and Definitions

3.1 A virtual world

THE INITIAL SEED FROM WHICH WATER FOOTPRINT approaches have
grown was planted in the early 1990s by Tony Allan, a prize-winning
expert on water and international relations, now Professor Emeritus
at King's College, London.[9] Allan was trying to understand why the
countries of the Middle East – one of the most hydrologically stressed
and geopolitically fractious places on the planet – had avoided war over
the region's shrinking water resources. He realised that they had resolved
the potentially crippling problems caused by a lack of water for irrigation
simply by importing agricultural produce. Food security in the Middle
East was therefore made possible by use of rainfall and rivers, lakes
and aquifers in other parts of the world. Allan coined the term 'virtual
water' to describe this (other researchers have since used alternative
terms such as embedded or shadow water to describe the same effect).

While Tony Allan was examining the implications of water use (or lack
thereof) in the Middle East, other researchers had begun to develop
methods for quantifying the impacts of human consumption on Earth's
biosphere. The first ecological footprints were published in the 1990s.[10]
These sought to equate multiple human demands for natural resources

to standardised units of measurement called Global Hectares. The calculations of ecological footprints took account of the land needed to produce the resources necessary to meet human demands; the space for accommodating buildings and roads which were essential to distribution of products; and impacts on ecosystems especially from waste and pollution. Ecological footprint calculations weren't able to satisfactorily account for water use though.

3.2 The numbers game

Into this gap stepped Arken Hoekstra, a professor at the University of Twente, and his PhD student, Ashok Chapagain (one of your authors). Hoekstra and Chapagain realised that it was possible to quantify the virtual water content of different products. From this, they began to calculate flows of virtual water between nations as a result of global trade. Based on these virtual water flows, they created a new measure of water use. The called this new measure the 'Water Footprint'.

Hoekstra and his team defined the Water Footprint as a spatially and temporally explicit indicator of freshwater consumption, measured over each phase of the production process and value chain. In other words, the Water Footprint of a product is the total volume of freshwater that is used to produce the goods and services consumed by an individual, company or nation.[11] In its simplest form, a Water Footprint is normally expressed in litres or cubic metres of water over a defined period of time.[12]

Initially all the Water Footprint accounts compiled by Hoekstra and Chapagain were based on publicly available data on hydrology, climate and agricultural water use.

How do you calculate the Water Footprint of a tomato?

Take a simple product like a tomato. To start calculating its Water Footprint you need to know a few key things about the conditions of the farm where it was grown. More specifically, two major types of factors are important:

- **When and where it was grown**: this will determine the rainfall, climatic conditions (temperature, humidity, exposure to sunshine hours, etc.), soil condition and suitability for tomato production.

- **How it was grown**: This will tell you whether it was irrigated or rain-fed, what type of irrigation was used, whether it was grown in under glass or in an open field, how much fertiliser, pesticide and other chemicals were used and so on.

Collecting this kind of data is often laborious work and quality and consistency of farm-scale data can be problematic. Happily, there are comprehensive and sufficiently reliable datasets existing which mean that researchers can short-cut the process. For instance, the UN Food and Agriculture Organisation houses a database of global crop production (FAOSTAT). So providing you know where your tomato was grown, researchers can, with a reasonable degree of confidence, tell you how you how much water has been used along its full supply chain, from the farm to your hands. Indeed, so calculations of this kind have now been undertaken that the Water Footprint Network now houses a global dataset of the Water Footprints for specific products, especially agricultural commodities.

3.3 **Size isn't everything**

If you want to understand your water-related impacts and risks, knowing the size of your Water Footprint is only the first step. But by itself this won't be enough information. As we discussed in our introduction, you also need to know more about where, when and by whom water is being used and how all of this affects economies, societies and ecosystems. The answers to such questions will reflect a variety of factors including:

- The demographic, socio-economic and cultural contexts in places where you have operations or supply chains.

- The degree of water scarcity and opportunity costs of water use in those places.

- The climate, geology and soil type where water is used for farming.

- The perspectives of multiple water users including those who are part of your operations and supply chain and those who are not.

- The location, design and operation of small- and large-scale water management infrastructure, e.g. irrigation ditches and storage reservoirs.

- The potential future competition over scarce freshwater resources.

- The likely impacts of climate change especially from more erratic rainfall patterns and increased variability in water availability.

- The vulnerability of river, lake, aquifers and wetland systems to water abstraction, pollution and other pressures.

- The extent and adequacy of water management policies, institutions and measures.

Many of these factors are difficult to quantify. The only way to really understand them is to travel to the parts of the world where water is being used in your operations or supply chains, listen to multiple stakeholders including water users and regulators, read the latest research on each topic and get to grips with the state of local water management and governance arrangements. For simple supply chains, and sophisticated companies, this prospect can be daunting. Once those chains become more complex and web-like, and for smaller companies which can devote fewer resources to the issue, the task can seem impossible.

Although it can't substitute for full understanding, Water Footprint Assessment can help.

3.4 Location, location, location . . . and colour

Water Footprints can be disaggregated into different components. These relate to the location where the water is actually used to make a product (not where the product is consumed); and to the source of water used to make the product (in simple terms, rivers or aquifers versus rainfall). Understanding the different components of a Water Footprint can aid understanding of specific parts of business operations or supply chains which are vulnerable to risk.

In terms of location, Water Footprints can be split into direct or indirect components; and internal or external use.

Your direct Water Footprint describes the volume of water that you use that comes 'from the tap'. For individuals, this is the same as water use in your home (assuming that you're one of the lucky minority on the planet that has access to reliable and potable domestic water supplies). In the

UK, for instance, we each use about 150 litres of water every day through a combination of taking showers, flushing toilets, filling kettles, rinsing and cooking food, laundering clothes, watering gardens and washing cars. This water will normally come from close to home. In the case of your authors, much of it is pumped from the chalk aquifers of Southeast England where we both live.

In the case of businesses, direct water use would include all the water used on site in factory or office locations owned and operated by your company, whether supplied by a utility or sourced from dedicated pumps or boreholes.

In contrast, your indirect Water Footprint incorporates all of that virtual water that Tony Allan described. It includes the water that farmers have used to grow the food that we eat, the irrigation for the cotton that made the shirt you're wearing, the cooling water that is lost to the sky as steam from the power station that keeps our lights on and the water that is used to make the dyes, or to lubricate the machines, or to clean the floors, in any one of a thousand factories which manufacture the stuff that we use in our day-to-day lives. If you're a fairly typical human being, your indirect Water Footprint will dwarf your direct water use. Staying in the UK, our average per capita Water Footprint has been estimated at more than 4500 litres a day – some 30 times the volume of our water use at home.[13]

The indirect Water Footprint of a business will include the water that was used to produce all the raw materials in the manufacturing process, the construction of buildings, the manufacture of equipment and the fuel and energy which powers factories, offices and vehicles throughout your supply chain. It can also include the water that is needed for consumers to use products such as laundry powder (for a consumer, the water used by a washing machine is part of his or her direct Water Footprint; for the manufacturer of the laundry powder, the same water counts as part of its indirect Water Footprint).

Water Footprints can also be broken into two broad types, or colours, of water use: blue or green. Green water refers to soil moisture used by rain-fed crops. The green Water Footprint of a product (normally food or cotton textiles) is equal to the volume of soil moisture lost through evapotranspiration during growth of rain-fed crops. Blue water refers to surface waters, such as rivers or lakes, and groundwater in the form of aquifers. Rivers, lakes and aquifers are the sources for almost all irrigation and for water used in industrial production and households. The blue Water Footprint of a product is therefore the amount of surface or ground water consumed in irrigated farms, in processing the products, and for cleaning and cooling purposes throughout the supply chain.

It's important to note that the volume of a blue Water Footprint is often smaller than the volume of water actually withdrawn from rivers, lakes or aquifers during the production process. This is because some water withdrawn from those water bodies by farms or factories may be returned in the form of unused or effluent flows, rather than being lost to the sky via evapotranspiration.

Finally, there is a third component of Water Footprint referred as grey Water Footprint (Figure 2, overleaf). The grey Water Footprint is a measure of pollution and is expressed as the volume of water required to assimilate the pollutant load to meet ambient water quality standards. The pollutant that requires the largest assimilation volume is used to calculate the grey Water Footprint. Grey Water Footprint Assessment is based on the quality and quantity of polluted effluent, and the volume and existing water quality of receiving water bodies such as rivers or lakes.[14] The biggest hurdle in estimating grey Water Footprints has been the difficulty in estimating the interaction of different pollutants with freshwater ecosystems and the resulting impacts on water quality. For these reasons grey water components have often been omitted from Water Footprint Assessments.

FIGURE 2. Different components of Water Footprint and water partitioning in the hydrological cycle.

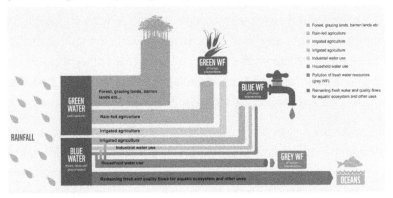

SOURCE: WWF and AfDB[15]

Understanding the Water Footprint of a tomato

On average it costs around 50 litres of water to grow one tomato weighing 250g. Out of this, about 25 litres is normally supplied through rainfall stored as soil moisture (i.e. through the green Water Footprint). Approximately 15 litres is the result of the evapotranspiration of irrigation water applied to the tomato plant (i.e. the blue Water Footprint). The remaining 10 litres is the grey Water Footprint arising from the fertilisers leaching to ground and surface water bodies. Note that these are all estimated global averages.

A study focused in tomato production in Spain revealed that here tomato production alone evaporates 297 million cubic meters of water in a year and pollutes another 29 million cubic meters.[16] The impact of global consumption of fresh tomatoes from Spain on Spanish freshwater resources depends heavily on local agricultural and climatic conditions, the status and on farm management of water resources and the total tomato production volumes and production system.

CHAPTER 4

Water Footprint Applications

SO FAR WE HAVE LEARNED ABOUT THE THEORY and basic approaches that underpin Water Footprint Assessment. But how can it be used in practice? The answer to this depends on who you are and what questions you're trying to answer.

To date, Water Footprints have been used in four basic types of assessments, summarised in Table 1 (overleaf). The first two types, product and business Water Footprint Assessments, have been led by companies or have very clear direct interest to businesses. The remaining applications, which focus on national and river basin scale Water Footprint Assessments, are of broader interest. If your business is concerned with practising good water stewardship – which by definition involves promoting efficient, equitable and sustainable management of water for wider social and economic benefit – you will need to understand the potential value of all of these approaches.

In this chapter we'll explain a bit more about each of these applications and give examples of how they have been used.

TABLE 1. Types of Water Footprint applications

	Type of Water Footprint Assessment	Typical questions addressed	Example data
Potential for direct business use	Product Water Footprint Assessment	What is the Water Footprint of a specific product? What is the Water Footprint of an individual who uses a range of products? What is the connection between product-specific and personal Water Footprints and ecosystems or other water users?	Water Footprint Assessments undertaken by SABMiller concluded that a litre of beer brewed in South Africa has a Water Footprint of 155 litres per litre of beer (47.1% net green water, 34.3% blue water, 18.6% grey water) and a litre of beer brewed in the Czech Republic has a Water Footprint of 45 litres per litre of beer (91.7% net green water, 5.9% blue water and 2.4% grey water).
	Business Water Footprint Assessment	Where and when in the value chain might a business potentially have adverse impacts on the environment and/or on other water users? What is the strategic water-related risk (reputational, physical, regulatory or other) to a business, and in which parts of the world does this risk arise? Where might be the priorities for action to manage water-related risk?	Based on the analysis of 1600 products representing 70% of its sales volume, Unilever found that 44% of its total domestic Water Footprint in water-scarce countries was associated with the manufacture and use of personal care products.

	Type of Water Footprint Assessment	Typical questions addressed	Example data
Potential for broader use to support water steward-ship	National Water Footprint Assessment	How reliant is the import and consumption of goods within a country on water resources within that country and elsewhere? What is the vulnerability of a nation to global water scarcity risks?	Of the UK's total agricultural Water Footprint, 62% is derived from virtual water embedded in imported products.
	River basin Water Footprint Assessment	What is the true picture of water stress or scarcity in a river basin over time? What are the economic implications of different water management scenarios in water-stressed basins?	The construction of 11 dams along the Lower Mekong River, and consequent impacts on freshwater fisheries, could result in a 4–7% overall increase in water use for food production throughout the Lower Mekong Basin, with a much higher estimated increase in Cambodia (29–64%) and Lao PDR (12–24%).

SOURCE: Adapted from Chapagain and Tickner, 2012

4.1 Product Water Footprint Assessments

As water issues have risen up the business agenda, many companies have sought to quantify risks and impacts of specific products and supply chains on freshwater ecosystems. One of the most popular and prominent uses of Water Footprint Assessment has involved putting a number on the volume of H_2O needed to grow, process or otherwise manufacture particular types of goods. The fact that it requires surprisingly large amounts of water to produce a cup of coffee (140 litres), a French baguette (155 litres), a cotton shirt (2500 litres) or leather products (17,000 litres per kilogram) has often been cited in newspapers and it is common to hear concerned politicians and captains of industry mention this kind of data in their speeches about water. Indeed, the results from these product Water Footprint Assessments have been instrumental in raising awareness of the links between consumption patterns, business risks and the global water crisis among a wide audience.

There is an important caveat to these otherwise compelling analyses. As we noted in our introduction, simply counting the number of litres it has taken to irrigate and process the wheat in a loaf of bread or the cotton in a shirt does not in itself tell us anything about the benefits or adverse impacts of business or consumer choices. As the study by SABMiller illustrated (see Table 1, previous pages), the same product – in their case a litre of beer – can have very different Water Footprints in different parts of the world depending on climatic conditions and on agricultural or industrial practices.

More importantly, the size of the Water Footprint does not necessarily dictate the impact of the product on freshwater ecosystems. A relatively small volume of water used in highly efficient irrigation schemes might result in an impressively low Water Footprint, but might still contribute to a harmful reduction in the flow of the river or stream that supplies the irrigation water. Conversely, a rain-fed crop grown in a temperate country might have a very large Water Footprint but, because it requires little or no irrigation, might exert negligible pressure on river flows. Local context is important.

Social considerations also play a part. It might be that poor farmers in some sourcing areas rely for their living on using water to irrigate 'thirsty' crops. That might mean that the end product a company makes has a high Water Footprint. It might also mean that river ecosystems are adversely affected by such heavy use of water. But if the alternative is that farmers and their families earn no money and go hungry, then using water may be – for the time being at least – the least-worst option.

The key message here is that the location and context of water use matter at least as much as the size of the Water Footprint.

The Water Footprint of asparagus production in Peru

Peru dominates the international trade in fresh asparagus and is the source for most of the asparagus sold in Australia, Europe and the USA. Asparagus production has become an important industry in the country, providing valuable jobs.

But delve deeper into the story and questions arise about the long-term sustainability of our asparagus consumption. Production is centred in the Ica Valley region, which receives less than 1 mm of rain a year. Asparagus, a thirsty crop, is being grown in one of the driest places on the planet. The farmers use cutting-edge water management practices, measuring every drop of irrigation and delivering to the root of the plants to minimise waste. Nevertheless, the ground water table in the region has declined by 1–3 meters since the asparagus boom started in 2002.

The multinational companies that run the asparagus farms have deployed bigger pumps so that they can continue to access ground water. But smaller-scale farmers who can't afford such kit are beginning to feel the hydrological pinch.

It's expensive for the big companies to run energy hungry water pumps and it's possible that market forces will intervene and they will need to find new ways, or new locations, to continue production. But it's likely that before this happens the poorer farmers and communities living in the Ica Valley will suffer from lack of access to water resources. Water scarcity is not the only factor we should consider in deciding whether or not asparagus production in Peru is a good thing. But it may become a more important issue if the asparagus industry, the Peruvian government and other local and international stakeholders can't find a solution to this problem.

For businesses or consumers used to grappling with the challenge of reducing carbon footprints, the complications involved in interpreting Water Footprint Assessments can be off-putting. Even organisations which specialise in water management, such as the Institution of Civil Engineers,[17] have fallen into the trap of presenting the challenge simply as one of reducing Water Footprints, when the reality is far more nuanced.

In order to address this complexity, some researchers have tried to simplify Water Footprint Assessment methods, drawing on methods developed in Life Cycle Assessments.[18] They have used measures of water stress in locations where products are grown, processed or manufactured to express product Water Footprints through a single quantifiable indicator. The intention here is to provide simple and communicable impact indicators, harmonised across products regardless of location through a process of weighting different variables. As we shall see in the following chapter, there remains considerable debate about the robustness of these weighted product Water Footprints.

Water Footprint Assessment, Life Cycle Assessment and ISO standards

The Water Footprint Network published a comprehensive manual on Water Footprint Assessment in 2011. *The Water Footprint Assessment Manual: Setting the Global Standard* aims to be the international standard regarding methodologies. It was developed internationally with involvement of a range of stakeholders. The methodology addresses impacts on water quantity and quality and guides users to go beyond numerical results and address the broader questions around sustainability, efficiency and equitability of water resources use.

More recently, the ISO published ISO 14046:2014 on Water Footprint.[19] This specifies principles, requirements and guidelines related to Water Footprint Assessment of products, processes and organisations based on Life Cycle Assessment (LCA) methods. It describes a framework for assessment and reporting but leaves the details of Water Footprint Assessment methodology open, arguing that 'there is currently no consensus on one accepted methodology for consistently and accurately associating inventory data with specific potential environmental impacts'. According to the ISO standard, 'the category indicator(s) and Water Footprint impact assessment method(s) shall be chosen based on the goal and scope of the study'. Nevertheless, the ISO standard describes two main impact categories to be considered in any comprehensive Water Footprint Assessment: i) impacts on water quantity; and ii) impacts on water quality. According to the ISO standard, the results of a Water Footprint Assessment can be expressed as a single value or a profile of impact indicator results.

The focus and scope of these approaches differs in a number of ways. The ISO standard focuses on local environmental impact assessment while Water Footprint Assessment (as set out in the WFN Manual) addresses mitigation of impacts too through, for instance, encouraging sustainable allocation of water resources. The ISO standard focuses specifically on the environmental impacts of products, while the methods in the WFN Manual provides guidance on the social perspective (equitability of water resources allocation) and the economic perspective (efficiency of water resources allocation) as well as environmental issues.

4.2 **Business Water Footprint Assessments**

Many companies, especially large multinationals, make or sell hundreds or even thousands of product lines. For them quantifying the Water Footprint of individual products is, if you can pardon the pun, a drop in the ocean. Such companies need to strategically manage key risks rather than just inventorise product Water Footprints. This involves analysing the role of water across the whole business and figuring out where the biggest water-related risks are across operations and supply chains that might span the globe.

It's not just FTSE 100 companies that can benefit from assessing risks in this way. Even small- or medium-sized outfits can find themselves at the whim of spikes in commodity prices caused by periodic drought in key locations or unforeseen shifts in water regulation which affect suppliers.

Given the fundamental role of water in their operations and supply chains, it is not surprising that large businesses in the food and beverages sector were the first to explore the potential for Water Footprint Assessment to aid understanding of company-wide risks. Companies such as SABMiller, The Coca-Cola Company, Nestlé and M&S began by conducting a range of specific product Water Footprint Assessments which showed them that significant elements of business impact and risk relating to water resources lay beyond the factory fence, especially in agricultural supply chains.

However, these companies quickly realised that simple volume-based measures of individual product Water Footprints inadequately reflected the complexity of water-related risks and would not be sufficient to inform targeted mitigation and management of strategic risks across their businesses. They needed more sophisticated tools. Happily for them, and largely as a result of NGO-led initiatives, such tools have recently emerged and continue to evolve.

TABLE 2. A summary of tools which help businesses undertake water-related risk assessment

	Ceres Aqua Gauge	GEMI Local Water Tool	WBCSD Global Water Tool	WFN WFA Tool	WRI Aqueduct	WWF/DEG Water Risk Filter	UN CEO Water Mandate Disclosure Guidelines	UN CEO Water Mandate Water Action Hub
Level of application	Portfolio	Facility	Portfolio	Facility/ Portfolio/ Product/ River Basin	Portfolio/ River Basin	Portfolio/ Facility	Portfolio	Portfolio/ Product
Physical, regulatory reputational water risk	✓	✓	Physical, reputation	Physical directly, regulatory and reputation inferred	✓	✓	✓	NA
Aggregates and quantifies risk	NA	✓		Quantifies water sustainability and efficiency	✓	✓	NA	NA
Format	Questionnaire	Excel	Excel, Interactive Tool	Interactive Tool	Interactive Tool	Interactive Tool	Questionnaire	Interactive Tool

	Ceres Aqua Gauge	GEMI Local Water Tool	WBCSD Global Water Tool	WFN WFA Tool	WRI Aqueduct	WWF/DEG Water Risk Filter	UN CEO Water Mandate Disclosure Guidelines	UN CEO Water Mandate Water Action Hub
Informs on future scenarios	NA		✓	✓	✓		NA	NA
Sector-specific Information		O&G	O&G	Agriculture, domestic, industry		34 industry sector		✓
Supports mitigation and response	✓	✓		✓		✓	✓	✓
Directly links to disclosure initiative	CDP Water, GRI	CDP, DJSI, IPIECA, GRI, Bloomberg	CDP, DJSI, IPIECA, GRI, Bloomberg	CEO Water Mandate Guidelines		CDP	Own	

For instance, WWF has worked with DEG (an agency of the German government) to develop a free-to-use online Water Risk Filter.[20] This has deployed by hundreds of companies to assess the water-related risk of thousands of business facilities across the globe. The World Resources Institute has separately produced a tool called Aqueduct[21] which claims to 'help companies, investors, governments, and other users understand where and how water risks and opportunities are emerging worldwide'. Each of these tools builds on, but goes significantly further than, Water Footprint Assessment. Table 2, on the previous page, briefly describes the most prominent tools which are available.

As a result it is now possible for SMEs, multinationals and investors in all sectors to undertake rapid high-level water risk assessments which can help them to target their risk management efforts. Some companies have gone even further. Puma, which makes sportswear, has sought to combine analysis of its water-related risks with a profit and loss methodology as part of an effort to measure, value and report environmental externalities (although it is not clear from Puma's literature whether this was achieved using standard Water Footprint Assessment approaches or through another method).

Partly as a result of a better understanding of water-related business risks, other companies have established risk-management initiatives that go beyond simple water efficiency or effluent treatment within factories. For example, The Coca-Cola Company, SABMiller, M&S, H&M and others have invested substantial resources in partnerships with NGOs and government agencies to try to address practical water resources management challenges.

4.3 National Water Footprint Assessments

There are already several ways in which governments and researchers measure water scarcity and water use at the national scale. Conventional country water accounts include data on withdrawal of blue water from rivers, lakes and aquifers for use in three main sectors: agriculture, industry and households. These accounts tend to draw on databases such as AQUASTAT,[22] owned by the UN Food and Agriculture Organisation, or The World's Water,[23] compiled by the Pacific Institute, a US-based think-tank.

These kinds of data adequately illustrate pressures on surface water and groundwater resources. But they provide limited information about the impacts and risks of water use within a country. They also say nothing about green water (i.e. rainfall) use in agriculture.

To help address these gaps, researchers have developed national, or sometimes regional, Water Footprint Assessments. The earliest attempt to assess the Water Footprint of different countries was made by Arjen Hoekstra (whom we met in Chapter 3) and his colleague P.Q. Hung in 2002. Hoekstra has since worked with various colleagues to publish several updates to this assessment.[24]

In brief, these studies sought to determine the Water Footprint of the products and services produced within and consumed by individual nations. The import and export of virtual water embedded in these products and services is taken into account to derive the Water Footprint of national consumption (Figure 3, overleaf). The total water used to produce goods and service in a country is known as Water Footprint of national production and includes products and services that might be exported, as well as those consumed domestically.

FIGURE 3. Components of national Water Footprint consumption and production.

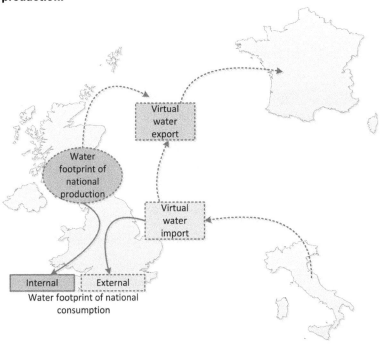

The process of calculating the volume of blue and green water used in different sectors across multiple countries is complex. Recent research has gone even further in search of accuracy, analysing climate and irrigation data at very fine spatial resolutions (to 5 by 5 arc minutes, equivalent to 10 km squares around the Equator) rather than using country average data.

Water Footprints of national consumption have been mostly calculated using what are called 'bottom-up' methods. These methods use the total consumption volumes of individual product, goods and services by the inhabitants of the country and the Water Footprint of producing these products estimated based on their origin.

Some recent work has used 'top-down' approaches which first calculate the total Water Footprint of national production (i.e. all the water used to produce goods and services within the country borders) and then use Multi-regional Input–Output tables (which show monetary flows between different economic sectors and countries) to trace the difference between production and consumption in an economy. Multi-regional Input–Output tables are still in development and are available for limited sectors of economy only, so at present this approach has limited utility.

4.4 River basin Water Footprint Assessments

If you are interested in water issues, there's a good chance that you will have seen maps depicting levels of water stress or scarcity in different countries. Typically, such maps are based on calculations of annual blue water withdrawals from rivers, lakes or aquifers as a proportion of rainfall and run-off. These calculations normally assume that all of the water taken from these natural sources is completely used, ignoring the fact that some water – such as that used to cool thermal power stations, for instance – is actually returned to rivers and can be re-used by downstream farmers, businesses or cities (water resource experts call this water 'return flows'). Very often the maps show average patterns of scarcity at the national scale, which can mask local variations in water availability, especially within large countries. Some illustrate the situation within specific river basins. And few of the calculations that underpin water scarcity maps take account of the amount of water that should remain in rivers, lakes or aquifers in order to sustain ecosystems and biodiversity.

To get a more accurate understanding of the pressure humans exert on rivers, lakes and aquifers, we need maps to take account of the water needed to keep ecosystems healthy, seasonal variations in rainfall and river flows and return flows from agricultural, industrial and municipal water uses. Water Footprint Assessment has enabled researchers to make these adjustments. In a paper published in 2012, by researchers from the University of Twente, the Water Footprint Network and others[25] analysed the blue Water Footprint of 405 major river basins around the world allowing for such factors. Their map, shown in Figure 4, looks

....

FIGURE 4. Blue water scarcity in 405 river basins between 1996 and 2005.

The darkest shading indicates river basins where more than 20% of water available in the basin is being used throughout the year. Some of these are in the most arid areas in the world (such as inland Australia) however other areas (such as western USA) have many months of water scarcity because significant amounts of water within these basins are being channelled into agriculture.

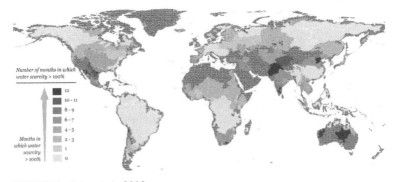

SOURCE: Hoekstra et al., 2012.

....

similar in some ways to conventional water scarcity maps: river basins such as the Indus in Pakistan or Lake Eyre in Australia all remain in darker shades. But there are differences. The Mississippi River in North America and the Ganges River in South Asia are also much darker in shades than on most conventional maps because in these basins the water use exceeds sustainable water availability in certain months of the year. Mississippi experiences "low to moderate" water scarcity for 4 months in a year, and Ganges experiences "significant to severe" water scarcity for 7 months in a year. The rather startling headline statistic from this study was that at least 2.7 billion people already live in river basins that experience severe water scarcity during at least one month of the year.

Beyond this global analysis, specific analyses of Water Footprints in particular river basins are rare. This is largely because the data on international trade and production that would be needed to calculate basin-scale Water Footprints are scarce (such data are normally available only at the national scale, although Arjen Hoekstra and his team have used production statistics at much finer detail to produce better Water Footprint numbers for crop and animal products). One exception is an assessment of potential shifts in land and Water Footprints in the Mekong Basin in Southeast Asia that was undertaken by Stuart Orr and others.[26] This research sought to understand potential implications of dam-building proposals on natural resources and local communities, especially through changes in availability of freshwater fish, a key source of protein in this part of the world. A different project in the Lake Naivasha basin in Kenya, also led by Orr, combined Water Footprint approaches with economic data to assess the value of water use in the horticulture and agriculture sectors.[27] This research concluded that there were marked differences between the socio-economic contribution of different water uses in the basin, in terms of dollar revenue and/or numbers of jobs per unit volume of water use.

....

CHAPTER 5

The Positives and Pitfalls of Water Footprints

AS EXPERIENCE OF USING Water Footprint applications has increased, so has our ability to assess their advantages and disadvantages. In the water research communities, there is an ongoing debate between those who are sceptical (such as Denis Wichelns, formerly of the International Water Management Institute, and David Zetland, an economist at Leiden University College in The Netherlands, who once authored a blog entitled 'Die Water Footprinting, Die'[28]) and others who are rather more enthusiastic (including Arjen Hoekstra and others). This debate sometimes spills over into the policy, business and water resource management fields too. It tends to focus on, and shift between, four different questions:

1. What are the best Water Footprint accounting standards?

2. How useful and responsible is it to use Water Framework Assessment results to communicate about water issues?

3. What value does Water Footprint Assessment have as a tool for business-risk assessment?

4. How can Water Footprint Assessment help inform better public policy for water management and economic development?

The first question is more of a methodological issue but the remaining three topics are more about the value of Water Footprint Assessment in the broader context of water resources management, business water use and public policy. We'll consider each of these topics in turn.

5.1 Water Footprint accounting standards

If recent conferences are to go by, an increasing number of researchers have become interested in Water Footprint Assessment. Given leaps in technology too, including better data storage, increased computing power and much more accurate remote sensing techniques, it's not surprising that the science has rapidly evolved since the late 1990s.

Improvements in the main climatic and hydrological databases on which Water Footprint calculations are based have been especially important. We can now take better account of local conditions. Data on flows of agricultural and other trades have also been updated, which is helpful.

There have also been conceptual leaps in this period. For instance, the idea of 'net green Water Footprint' has been suggested as a means of distinguishing between the green Water Footprint of a planted crop and that of the natural or semi-natural vegetation that would have existed if farmers hadn't cultivated the land.[29] In theory, this helps us understand whether rain-fed agriculture is having more of an impact on soil moisture than nature would have if left (more or less) undisturbed. But is also controversial given that it might be impossible to know what natural vegetation would have looked like on land which has been cultivated for decades or centuries.

In an attempt to consolidate analytical approaches, the Water Footprint Network published the *Water Footprint Assessment Manual*.[30] This sets out robust and standardised methods for calculating volumetric Water Footprint accounts and provides guidance on how to interpret them given variations in context.

Despite the efforts of the Water Footprint Network, some methodological aspects have yet to be fully resolved. For example, though there is a standardised approach to grey Water Footprint accounting, the absence of necessary minimum water quality data stops a complete sustainability assessment of grey Water Footprint. By definition, grey water remains within the local river, lake or aquifer system, ready to be used by people repeatedly providing it is treated to a sufficient level. Whether or not to treat it as 'lost' from the local system, in the same way that blue or green water is considered lost once it has evapotranspired from fields into the atmosphere, is a tricky conceptual question.

More practically, the varying water quality standards in different parts of the world make it difficult to account for grey Water Footprint. Hypothetically, Country A might have a policy stating that the maximum concentration of a given pollutant should be 100 parts per million; but it's neighbour, Country B (which might share the same river) might have a standard that puts the limit at only 10 parts per million. If factories on either side of the border emit the same amount and type of pollution (i.e. the same grey water), which figure should be used to calculate the total grey Water Footprint being exerted on the river? The Water Footprint Network has provided guidelines on how to address such situations but ultimately it will be for the governments or water management agencies of the respective countries to agree a way forward.

The other significant accounting Water Footprint challenge is linked to the concept of weighted Water Footprints. One potential merit of a standard

Water Footprint account is that it enables water users to discover which component of their Water Footprint they should be most concerned about (e.g. blue, green, grey; direct, indirect). It follows that interpretation of Water Footprint accounts and what they might imply for companies, governments, water users and ecosystems is a complex business.

Throw in issues relating to environmental, social, economic and political factors where the water is originally used and you can see that understanding the implications for any single stakeholder is as much art as science and there are unlikely to be binary thresholds. Anyone interpreting a Water Footprint Assessment needs to consider not only environmental factors relating to blue water resources (rivers, lakes, aquifers) but also land tenure, land allocation and social equity issues linked to green water (i.e. soil moisture). These are thorny topics.

Weighted Water Footprint assessment is an attempt to reduce all of this complexity. Borrowing from Life Cycle Assessment approaches, weighting factors have mostly reflected on different indicators of environmental stress such as water scarcity. But social and economic factors – the opportunity costs of water uses, inequalities of access to water, and so on – are far more difficult to incorporate and most can be weighted only in a qualitative way. So what is meant to be a simple and objective accounting approach might mask some important complexities and might be, in part, based on subjective analysis.

As we've learned, Water Footprint accounting has attracted vociferous criticism. Many of those who are sceptical have, correctly, highlighted the danger that a simple volumetric Water Footprint indicator (litres per loaf of bread, say) might be misunderstood to be a proxy for the actual impacts of water use. Using weighted Water Footprints to simplify the picture increases that risk.

5.2 Water Footprint Assessment as a communications tool

Every three years, the UN publishes its World Water Development Report, a comprehensive survey of the water-related challenges facing the world. In recent years, a theme of these reports has been the need for water sector professionals to communicate more with those outside of what the UN calls 'the water box'. In other words, water experts have to get much better at engaging decision-makers in government, the private sector and civil society so that economic and social development becomes more hydrologically sustainable.[31]

If this is to happen, water experts have to find stories, data and language that will resonate more with decision-makers than their standard jargon (such as Integrated Water Resources Management). Experience suggests that Water Footprint Assessment could be helpful here. Newspapers such as *The Guardian* have put the results of Water Footprint assessment of the UK on their front pages.[32] Anecdotal evidence from our conversations with a variety of audiences tells us that volumetric Water Footprint statistics have been a powerful way of getting people to think about water in a very different way. Eye-opening statistics about our true reliance on water for the stuff we wear, eat and consume every day have helped at least some decision-makers to understand how global water security underpins our lifestyles and economies.

But care is needed. As we discussed in Chapter 3, you can't understand impacts and risks merely from a volumetric Water Footprint. Indeed, stories about how much virtual water we use in the form of imported goods (i.e. our indirect Water Footprint) might give the impression, unintentionally, that we the consumers are to blame for water scarcity, poverty and ecosystem collapse in, say, an exporting country in Africa.

In truth, the water problems in that exporting country are likely to be complex and implying that consumers are somehow to blame may cause more harm than good if it reduces demand for important export crops or other goods. This is the main reason why there is general scepticism about the use of Water Footprint labels on consumer goods.

5.3 Water Footprint Assessment as a business-risk tool

There's little doubt that Water Footprint Assessment and related applications can be useful tools for companies who want to map and understand water-related business risks and impacts. The number of businesses which have used it, often in partnership with NGOs and researchers, has grown rapidly tool in the last ten years. With each attempt, new lessons are learned.

For instance, M&S (a UK-based food and clothing retailer) originally (in 2007) set out to assess the Water Footprint of all of the food products on its shelves. Having realised how much effort this would take, the company worked with WWF to deploy the Water Risk Filter tool instead. This provided a more strategic analysis of immediate priorities risk mitigation investments.

The experience of SABMiller (a large global brewing company) showed how Water Footprint Assessment can help managers within the company to understand the nature of water risks, and especially to get a grip on the fact that most risks lie beyond the factory fence. In the case of SAB Ltd (the company's South African subsidiary) an analysis illustrated that 98.3% of the total Water Footprint was related to crop growth in South Africa. This helped to motivate SAB Ltd to develop partnerships to address water scarcity issues in the Gouritz Water Management Area where the company's hop supply chain is concentrated.

While Water Footprint Assessment can help companies to understand water-related risks, it offers no information on potential courses of action to address these risks. Other tools and guidance have been developed to help companies understand how to engage with a range of stakeholders in order to promote and resource context-specific water management initiatives that can benefit communities and ecosystems as well as reducing business risks.[33]

5.4 Public policy in light of Water Footprint Assessment

Despite much debate in conference halls and academic papers, and despite the proliferation of national Water Footprint assessments, there have been few examples to date of Water Footprint Assessment influencing national water policies. Perhaps the only example where the Water Footprint concept has gained clear policy traction is in Spain where the national government requires any new development project to undertake a Water Footprint Assessment, linked to the preparation of river basin management plans as required by the EU Water Framework Directive.[34] In the UK, government departments and agencies have shown interest but have gone no further than commissioning research.

The limited influence of Water Footprint at the policy level may be simply because it is a new tool and there hasn't been sufficient time for it to filter into policy-making. In addition, in many countries where water resources are under pressure water policy has already been amended to set a framework for better management. In such countries, it may not be clear to policy-makers what the added benefit is of undertaking Water Footprint Assessments when they already know well the degree to which their rivers and aquifers are overstretched. The worry here is that putting a focus on Water Footprint Assessment may distract from efforts to implement recent water policy measures.

Water Footprint Assessment may yet play a role in informing economic policy in some parts of the world. A fascinating analysis was undertaken of economic water-related risks around Lake Naivasha, Kenya.[35] This work combined Water Footprint data with economic tools to assess the implications of water scarcity for economic development scenarios, especially as they relate to agriculture, energy production and trade. The idea was to illustrate opportunities to deliver economic development which could be within the limits of hydrological sustainability and to illustrate some of the medium- to long-term social and economic consequences of using water resources unsustainably. The author of this report, Guy Pegram, has described it as a 'water in the economy' analytical approach, distinguishable from the 'water for the economy' route conventionally taken by water managers who historically have concerned themselves with enhancing water-supply options in response to economic development.

Some commentators, including Arjen Hoekstra, have gone even further and suggested that Water Footprint analyses point to the need to amend international trade frameworks such that they facilitate more hydrologically sustainable global consumption patterns.[36] This is absolutely correct on one level: current frameworks governing trade and development pay lip service to environmental concerns, including those linked to water resources. But it seems unlikely for now that the World Trade Organization or other institutions will radically alter their ways.

....

CHAPTER 6

Conclusions

WATER FOOTPRINT ASSESSMENT can be useful. It is potentially helpful for business analysts, communications experts and policy-makers as much as for water experts. Much of the value of Water Footprint tools stem from their ability to inform analyses of consequences and constraints of economic activity linked to water resources and freshwater ecosystems.

Most tools are designed to help with specific tasks. Water Footprint Assessment is no exception. It has helped to improve understanding of the hydrological interdependence of nations, linked to international trade. As a metaphor and as a set of metrics, Water Footprint has proven to be a powerful aid to communication 'outside the water box' to a range of policy-makers and the public – although this remains work in progress. It has been especially effective for strategic business risk assessment and for motivating CEO and board-level engagement.

A Water Footprint Assessment is not a silver bullet though. Methodologies are evolving. Other tools are often required to address water-related risks. Responses to Water Footprint data need to take account of broader environmental, social and economic issues, which are dictated by local context.

In the box below we have suggested a set of 'golden rules' that might be helpful to guide the deployment of Water Footprint tools in the water management, business and research communities. If you go about it thoughtfully, you can use Water Footprint to shed light on the deep and sometimes surprising connections between water and business risk, trade patterns, supply chains, ecosystem health and social equity.

Some golden rules for Water Footprint Assessment

- Water Footprint Assessment has already led to better understanding of the true picture of water scarcity at the global, regional and national levels and of the link between water security challenges and food security. Researchers should prioritise the development and refinement of applied tools for use by businesses, policy-makers and stakeholders as they seek to address such challenges. Further critical examination of the response of the businesses to risks highlighted in Water Footprint Assessments is needed.

- Water Footprint Assessments should be used by stakeholders for communication to wider audiences beyond the expert community. Scientific analysis, and visual tools such as virtual water flow maps, can quickly build awareness of the links between global water security, trade, national food or energy security and social or economic development goals.

- Such communication should be carefully planned. An emphasis on the total volumetric Water Footprint indicator can grab the headlines, but without explanation and contextualisation it can lead to misinterpretation of the issue and, potentially, to perverse responses. Messaging should incorporate at least a minimum of explanation about

---

the environmental impacts of Water Footprints and about social and economic benefits of water use.

- Because of this need for contextualisation, extreme care should be taken in interpretation of weighted Water Footprint data. Similarly, use of 'single number' Water Footprint indicators to label consumer goods is seldom appropriate or helpful and businesses should be wary of their use.

- Businesses should use Water Footprint Assessment to understand strategic corporate risk related to water. There is a range of Water Footprint tools available to guide companies in this endeavour. Many are freely available online and increasingly straightforward to use. Companies should also draw on tools which provide guidance on appropriate responses to Water Footprint-based risk assessments, including those which advise on stakeholder engagement, partnership development and dialogue with governments about public policy.

- Governments and the research community should further explore the use of Water Footprint tools to improve understanding of the links between water resources and economic development planning. Combining Water Footprint applications with economic and social analytical tools has the potential to aid development of more sophisticated and hydrologically sustainable options in pursuit of social and economic outcomes.

- Although Water Footprint Assessment may not be useful to stimulate development of new public policy on water management, governments and other stakeholders (such as NGOs) should target communication of Water Footprint data to encourage involvement of new political and economic actors in dialogue to support implementation of existing water policy. For instance, a broader cross-section of the business community might be motivated to contribute to water resources management efforts, including hydrological monitoring, if they are shown data that link supply chain risks to increasing water scarcity or pollution in catchments from which they source goods or services.

- Use of Water Footprint data to call for more sustainable global trade and economic development frameworks may be helpful but calls to action should take account of (or at least acknowledge the need to take account of) a wider set of environmental, social and economic issues linked to water use. They should also recognise the necessity and inevitability of political trade-offs between these issues.

- Any organisation considering undertaking a Water Footprint Assessment should be clear on why it is doing so, and on how it will use the results. The purpose might be exploratory (to understand Water Footprint approaches and to help decide whether they can be useful, for instance) or applied. A Water Footprint Assessment for its own sake is seldom useful.

SOURCE: Adapted from Chapagain and Tickner, 2012[37]

Notes and References

Further reading

THE WATER FOOTPRINT NETWORK WEBSITE has a comprehensive set *of case studies,* papers and guidance notes on Water Footprint Assessment, including *The Water Footprint Manual. The Water Footprint Manual* sets out more comprehensive guidance on methods for Water Footprint Assessment: Hoekstra, A.Y., Chapagain, A.K., Aldaya, M.M. and Mekonnen, M.M. 2011. *The Water Footprint Assessment Manual: Setting the Global Standard* (London: Earthscan), **http://waterfootprint. org/en/resources/publications/**

For those interested in issues relating to business water risks, you can read more about mitigation options for businesses facing water-related risk in Orr, S. and Pegram, G. 2014. *Business Strategy for Water Challenges: From Risk to Opportunity* (Oxford: Dō Sustainability), **http:// www.dosustainability.com/shop/business-strategy-for-water-challenges-from-risk-to-opportunity-p-57.html**

Various tools are available for those interested in assessing business risks relating to water, many of which incorporate or are adapted from Water Footprint Assessment. These include the following:

- The Water Footprint Network's Water Footprint Assessment tool: **http://waterfootprint.org/en/resources/interactive-tools/water-footprint-assessment-tool/**

- The World Business Council for Sustainable Development's Water tool: http://www.wbcsd.org/work-program/sector-projects/water/global-water-tool.aspx

- The WWF-DEG Water Risk Filter: http://waterriskfilter.panda.org/

- The World Resources Institute's Aqueduct tool: http://www.wri.org/our-work/project/aqueduct

- The Aqua Gauge tool developed by Ceres: http://www.ceres.org/issues/water/corporate-water-stewardship/aqua-gauge/aqua-gauge

- The GEMI Local Water Tool: http://gemi.org/localwatertool/

Endnotes

1. For a comprehensive assessment of global water-related challenges, read the UN World Water Development Report published in 2012: http://www.unesco.org/new/en/natural-sciences/environment/water/wwap/wwdr/wwdr4-2012/

2. The Pacific Institute, a US think-tank, compiles a water conflict map and chronology which is available online at http://www2.worldwater.org/conflict/map/

3. *The Guardian*, http://www.theguardian.com/world/2014/apr/10/drought-brazil-coffee-beans-prices

4. For instance, WWF produces a biannual Living Planet Index which tracks the state of ecosystems and biodiversity. The October 2014 LPI showed that populations of vertebrate species living freshwater ecosystems around the world had declined by 76% since 1970.

5. World Economic Forum. 2014. *Global Risks Perception Survey 2014* (Geneva: World Economic Forum).

6. See Lloyds & WWF. 2010. *Global Water Scarcity: Risks and Challenges to Business* (London: Lloyds) at http://www.lloyds.com/news-and-insight/news-and-features/environment/environment-2010/lloyds_report_highlights_water_scarcity_threat

7. See https://www.cdp.net/CDPResults/CDP-Global-Water-Report-2013.pdf

8. Water stewardship can be defined as a progression of increased improvement of water use and a reduction in the water-related impacts of internal and value chain operations. More importantly, it is a commitment to the sustainable management of shared water resources in the public interest through collective action with other businesses, governments, NGOs and communities. It is distinct from, but builds on, water efficiency because it obliges a company to move beyond its factory fence to engage other stakeholders in the river basin or territory where its risk is located.

9. Tony Allan was awarded Stockholm World Water Prize in 2008, the Florence Monito Water Prize in 2013 and the Monaco Water Prize in 2013 for his work on virtual water. He summarised his thinking in a book: Allan, T. 2011. *Virtual Water: Tackling the Threat to Our Planet's Most Precious Resource* (London: I.B. Tauris).

10. See Rees, W.E. and Wackernagel, M. 1994. Ecological Footprints and Appropriated Carrying Capacity: Measuring the Natural Capital Requirements of the Human Economy. In *Investing in Natural Capital: The Ecological Economics Approach to Sustainability* (eds A.M. Jansson, M. Hammer Folke and R. Costanza), pp. 362–390 (Washington, DC: Island Press).

11. This is the definition set out in Hoekstra, A.Y., Chapagain, A.K., Aldaya, M.M. and Mekonnen, M.M. 2011. *The Water Footprint Assessment Manual: Setting the Global Standard* (London: Earthscan).

12. In most parts of the world, cubic metres – or cumecs (m^3) – are the standard unit of measurement used by water resource measures (the notable exception is the USA where acre feet is often used).

13. Chapagain, A.K. and Orr, S. 2008. *UK Water Footprint: The Impact of the UK's Food and Fibre Consumption on Global Water Resources* (Godalming, UK: WWF-UK).

14. Franke, N.A., Boyacioglu, H. and Hoekstra, A.Y. 2013. *Grey Water Footprint Accounting: Tier 1 Supporting Guidelines* (Delft, The Netherlands: UNESCO-IHE).

15. WWF and African Development Bank. 2012. *Africa Ecological Footprint Report: Green Infrastructure for Africa's Ecological Security* (Gland, Switzerland: WWF International).

16. Chapagain, A. K. and Orr, S. 2009. An improved Water Footprint methodology linking global consumption to local water resources: A case of Spanish tomatoes. *Journal of Environmental Management* (Volume 90): 1219–1228.

17. See Institution of Civil Engineers. 2010. *Global Water Security: An Engineering Perspective* (London: The Royal Academy of Engineering).

18. For instance, see Ridoutt, B.G. and Pfister, S. 2009. A revised approach to Water Footprinting to make transparent the impacts of consumption and production on global freshwater scarcity. *Global Environmental Change* (Volume 20): 113–120.

19. ISO. 2014. *ISO 14046:2014 Environmental management – Water Footprint – Principles, Requirements and Guidelines* [Geneva: ISO].

20. See http://waterriskfilter.panda.org/

21. See http://www.wri.org/our-work/project/aqueduct

22. FAO. 2011. *AQUASTAT online database*, http://www.fao.org/nr/water/aquastat/data/query/index.html

23. Pacific Institute. 2011. *Water Data from The World's Water*, http://www.worldwater.org/data.html

24. Key papers on national Water Footprint Assessment include: Hoekstra, A.Y. and Hung, P.Q. 2002. Virtual water trade: A quantification of virtual water flows between nations in relation to international crop trade. UNESCO-IHE, Delft, The

Netherlands; Chapagain, A.K. and Hoekstra, A.Y. 2004. Water Footprints of nations. UNESCO-IHE, Delft, The Netherlands.

25. Hoekstra, A.Y., Mekonnen, M.M., Chapagain, A.K., Mathews, R.E. & Richter, B.D. 2012. Global monthly water scarcity: Blue Water Footprints versus blue water availability. *PLoS ONE* (Volume 7): e32688–e32688.

26. Orr, S., Pittock, J., Chapagain, A. and Dumaresq, D. 2012. Dams on the Mekong River: Lost fish protein and the implications for land and water resources. *Global Environmental Change* (Volume 22): 925–932.

27. Pegram, G. 2010. Shared risk and opportunity in water resources: Seeking a sustainable future for Lake Naivasha. WWF International & PEGASYS, Gland, Switzerland.

28. You can read David Zetland's blog on Water Footprinting at **http://www. aguanomics.com/**. It's embedded into his post dated 2 May 2012.

29. See **http://www.sab.co.za/sablimited/action/media/downloadFile?media_ fileid=918**

30. Hoekstra, A.Y., Chapagain, A.K., Aldaya, M.M. and Mekonnen, M.M. 2011. *The Water Footprint Assessment Manual: Setting the Global Standard* (London: Earthscan).

31. World Water Assessment Programme. 2009. *The United Nations World Water Development Report 3: Water in a Changing World* (Paris: UNESCO/London: Earthscan).

32. See **http://www.theguardian.com/environment/2008/aug/20/water.food1**

33. You can read more about mitigation options for businesses facing water-related risk in Orr, S. and Pegram, G. 2014. *Business Strategy for Water Challenges: From Risk to Opportunity* (Oxford: Dō Sustainability), **http://www.dosustainability.com/ shop/business-strategy-for-water-challenges-from-risk-to-opportunity-p-57.html**

34. Aldaya, M.M., Garrido, A., Llamas, M.R., Varelo-Ortega, C., Novo, P. and Casado, R.R. 2010. *Water Footprint and Virtual Water Trade in Spain. Water Policy in Spain* (Leiden, The Netherlands: CRC Press), pp. 49–59.

35. Pegram, G. 2010. Shared risk and opportunity in water resources: Seeking a sustainable future for Lake Naivasha. WWF and PEGASYS, Godalming, UK.

36. See Hoekstra, A.Y. 2006. The global dimension of water governance: Nine reasons for global arrangements in order to cope with local water problems. Value of Water Research Report Series No. 20, UNESCO-IHE, Delft, The Netherlands.

37. Chapagain, A.K. and Tickner, D. 2012. Water footprint: Help or hindrance. *Water Alternatives* (Volume 5): 563–581.

....

For Product Safety Concerns and Information please contact our EU
representative GPSR@taylorandfrancis.com
Taylor & Francis Verlag GmbH, Kaufingerstraße 24, 80331 München, Germany